A BOOK IN EVERY HOME

Containing Three Subjects:

Ed's Sweet Sixteen,
Domestic and
Political Views.

By

EDWARD   LEEDSKALNIN

Martino Publishing
Mansfield Centre, CT
2012

Martino Publishing
P.O. Box 373,
Mansfield Centre, CT 06250 USA

www.martinopublishing.com

ISBN   978-1-61427-350-9

© 2012 Martino Publishing

Cover design by T. Matarazzo

Printed in the United States of America On 100% Acid-Free Paper

A BOOK IN EVERY HOME

Containing Three Subjects:

Ed's Sweet Sixteen,
Domestic and
Political Views.

By

EDWARD LEEDSKALNIN

Published by

EDWARD LEEDSKALNIN

Homestead, Florida

Preface

The Coral Castle of Florida is one of South Florida's historical and famous landmarks. The story of this beautiful and fascinating wonderland of coral rock reads like a fairy tale, and began over fifty years ago when an obscure Latvian immigrant named Edward Leedskalnin came to South Florida. He chose this sparsely settled section of Florida because he "wanted to get away from the world". Here he began many years of self-induced hard labor in order to forget his unrequited love for his one and only sweetheart. This young girl, whom he always referred to as his "Sweet Sixteen", jilted him for another on the eve of their wedding day in his native Latvia. An extremely sensitive soul, he was deeply hurt and made the decision to leave his native country and seek solace elsewhere. Emigrating to America in 1921, he went directly to California, staying there but a few months, and then moving on to Texas. From Texas he made his way to Florida. He did not stop until he came within sight of the very southernmost tip of Florida, south of Florida City.

Here on the edge of the Everglades he apparently felt he was far enough away from people, and settled

down on a small plot of land which a generous neighbor permitted him to use. He built a house from logs cut from the surrounding pine trees, and coral rock which he quarried on the premises. Thus began the strange and unusual project he set for himself, interrupted only by his moving to a larger plot of ground three miles north of Homestead, which after twenty-five years of continuous labor culminated in the completion of what can only be described as a "Coral Castle", breathtaking in scope and imagination, unbelievable for skill and patience required.

Today, the Coral Castle of Florida, as it is known, lies athwart U. S. Highway No.1 about twenty-five miles south of Miami. The Coral Castle is no ordinary structure. It is set on a ten-acre tract of land, the castle proper being surrounded by an eight-foot high wall made of huge blocks of coral rock, each weighing several tons. The tower contains 160 tons of coral rock, each block making up this huge building weighing nine tons, the first floor of which Leedskalnin used as a workshop and the second floor housed his living quarters. An air of mystery prevailed about these quarters since no one was permitted entry. Behind the huge walls, in beautiful settings, are the most marvelous and fantastic pieces of coral rock furniture, and movable objects which

he created from his fertile imagination. There are rocking chairs weighing thousands of pounds so delicately balanced they move at the touch of a finger. Couches, beds, chairs, tables of all sizes and shapes, including one table hewn from solid coral rock into the shape of the State of Florida, and another cut into the shape of a heart with a beautiful ever blooming floral centerpiece growing out of the center of this rock table. There are huge crescents atop walls twenty feet high, an obelisk reaching up to the sky weighing twenty-eight tons that Leedskalnin set in place by the use of simple hand tools. There is an ingenious polaris telescope carved out of the coral rock rising twenty-five feet into the air. With this telescope he would study astronomy nightly under the Florida skies. As in most castles, there is a subterranean well, with a circular staircase carved out of the rock leading down into the water.

Upon entering the Coral Castle it is necessary to go through the swinging gate, a triangular block of stone weighing three tons that turns, permitting you to pass in. In the center rear wall is a huge stone gate weighing nine tons reminiscent of ancient castles that is so perfectly pivoted and balanced that a child can turn it. On the opposite side is constructed an iron ladder

which allows you to get up onto the huge Crescent of the East, from which one can overlook the entire park.

Leedskalnin never forgot his "Sweet Sixteen", and much of the architecture he created was designed with the thought uppermost in his mind that some day she would return to him. This is apparent from the complete bedroom hewn out of the coral rock furnished with twin beds, a pair of children's beds, as well as a baby cradle; also from the beautiful patio for children to play in, called the "Grotto of the Three Bears", containing all the furniture in the famous fairy tale, including a huge coral porridge bowl. He never married, living the life of a true hermit in this coral wonderland of his.

The Coral Castle has become a familiar sight over the years to travelers on U. S. Highway No. 1, and many, intrigued by this huge coral edifice, have taken the time to explore it, while others raced by wondering at its incongruity in this modern world of ours. Those who stopped were met by the man who constructed this monumental work, and guided around while he explained its many mysteries. But one thing he never told anyone was how he ever was able to move the huge coral rocks weighing up to thirty-five tons which he excavated single-handed. When asked, he replied simply that he knew the secrets used in

the building of the Pyramids of Egypt. Whatever the secrets or principles of construction he used, they died with him when he passed away in a Miami hospital in December 1951. It can be said his knowledge of the practical use of pulleys, and leverage was not exceeded by the ancient Greeks or Egyptians.

Edward Leedskalnin's remarkable versatility is shown also by his studies and experiments in magnetic current, and he evolved and published many theories on the subject which he claimed are more accurate than any of the known thoughts on magnetism. For instance, his theory of Cosmic Force is remarkably akin to Einstein's latest theory of Unity. In the pamphlet he published in 1946, he states, "the North and South poles are the cosmic force. They hold together the earth and everything on it—turning the earth around on its axis".

In an article reprinted in the Miami Daily News, he writes, "Every form of existence, whether it be rock, plant or animal life, has a beginning and an end, but the three things that all matter is constructed from has no beginning and no end. They are the North and South poles' individual magnets, and the netural particles of matter. These three thing are the construction blocks of everything". In addition to all

these studies, he found time to write and publish this booklet called, "A Book in Every Home", giving his opinions on life in general based on his observations of people. Here one can gain an insight to Leedskalnin's ideas as he states, "An educated person is one whose senses are refined. We are born as brutes, we remain and die as the same if we do not become polished. But all senses do not take polish. Some are too coarse to take it. The main base of education is one's 'self-respect'. Anyone lacking self-respect cannot be educated. The main bases of self-respect is the willingness to learn, to do only the things that are good and right, to believe only in the things that can be proved, to possess appreciation and self-control".

The Coral Castle is known to countless thousands, and is considered one of the wonders of the world, baffling engineers and laymen alike. A masterpiece of rugged and permanent beauty, it is a landmark of the Redlands district of South Florida, and Edward Leedskalnin himself has become a legendary figure. Compared by many authorities to the Stonehenge in England, and the Pyramids of Egypt, Coral Castle of Florida is now the mecca for visitors from all over the world. Upon seeing the Coral Castle, a prominent minister stated, "After seeing this, I can give a great sermon on the quality of patience". A

noted visitor, Dr. David Fairchild, founder of Fair-
child Tropical Gardens, wrote in the guest book,
"This is one of the most interesting spots in Florida.
I am glad I knew Edward Leedskalnin".

Thus, although he never again saw the girl who
inspired him, the Coral Castle of Florida stands now
and forever as a monument for all to see and marvel
at the genius, imagination, and skill of this strange
but brilliant man, Edward Leedskalnin.

AUTHOR'S PREFACE

Reader, if for any reason you do not
like the things I say in this little book, I
left just as much space as I used, so you
can write your own opinion opposite it
and see if you can do better.

The Author

Ed's Sweet Sixteen

To those more than fifteen thousand people who have seen Ed's Place, I told about Ed's Sweet Sixteen. Now, I will tell you why I did not get the girl.

In Ed's Place, there was a lasting fame for a girl's name but it would have taken money to put the fame upon her. The trouble was that I did not have the money and did not make enough. That was the reason I could not look for a girl.

Now, I am going to tell you what I mean when I say "Ed's Sweet Sixteen". I don't mean a sixteen year old girl, I mean a brand new one. If it had meant a sixteen year old girl, it would have meant at the same time, that I made money for the sweet sixteen while she was making love with a fresh boy.

I will furnish all the love making to my girl. She will never have to seek any from anybody else, for I believe that there

1

is not a boy or a man in this world good enough to be around my girl and I believe that the other men also ought to have enough self-respect so that they would think that I am not good enough to be around theirs.

Anything that we do leaves its effect, but it leaves more effect upon a girl than it does upon a boy or a man, because the girl's body, mind and all her constitition is more tender and so it leaves more impressions— and why should one want to be around any-body's else impressions?

A girl is to a fellow the best thing in this world, but to have the best one second hand, it is humiliating.

All girls below sixteen should be brand new. If a girl below sixteen cannot be called a brand new any more, it is not the girl's fault; the mamma is to blame! It is the mamma's duty to supervise the girl to keep those fresh boys away.

In case a girl's mamma thinks that there is a boy somewhere who needs exper-

ience, then she, herself, could pose as an experimental station for that fresh boy to practice on and so save the girl. Nothing can hurt her any more. She has already gone through all the experience that can be gone through and so in her case, it would be all right.

But all the blame does not rest on the mamma alone. The schools and the churches are cheapening the girls! They are arranging picnics—are coupling up the girls with the fresh boys—and then they send them out to the woods, parks, beaches and other places so that they can practice in first degree love making.

Now, I will tell you what the first degree love making is. The first degree love making is when the fresh boy begins to soil the girl by patting, rubbing and squeezing her. They start it in that way but soon it begins to get dull and there is no kick in it, so they have to start in on the second degree and keep on and then by and by, when

the right man comes along and when he touches the girl, then he touches her like dead flesh. There is no more response in it because all the response has been worked out with those fresh boys. Why should it be that way?

Everything we do should be for some good purpose but as everybody knows there is nothing good that can come to a girl from a fresh boy. When a girl is sixteen or seventeen years old, she is as good as she ever will be, but when a boy is sixteen years old, he is then fresher than in all his stages of development. He is then not big enough to work but he is too big to be kept in a nursery and then to allow such a fresh thing to soil a girl—it could not work on my girl. Now I will tell you about soiling. Anything that is done, if it is with the right party it is all right, but when it is with the wrong party, it is soiling, and concerning those fresh boys with the girls, it is wrong every time.

Now, how can you find out if I am right? Pick out any girl you want but do it before she has anything to do with anybody—as soon as she begins to couple herself with somebody. You watch her every day and some day you will see the girl coming home with a red face. One's face is a window for other people to look in on and when it turns red it shows that there was something done that her moral conscience told her should not have been done.

It is shocking to imagine that someone else produced that red face to my girl. In such a case she could not be one hundred per cent sweet. As soon as a girl acquires experience the sweetness begins to leave her right away. The first experience in everything is the most impressive. It should be reserved for the permanent partner—the less of the new experience is left, the cheaper the affair will be.

That is the reason why I want a girl the way Mother nature puts her out. This

means before anybody has had any chance to be around her and before she begins to misrepresent herself. I want to pick out the girl while she is guided by the instinct alone.

When I started out in doing things that would make it possible for me to get a girl, I set a rule in my behaviour to follow:

The sweet sixteen had to be a beginner and a likeable girl and with a mild disposition; I had to be deserving of her. Everybody's sweet sixteen should be so high in one's estimation that no temptation could induce one to act behind her back. I always have wanted a girl but I never had one.

The reason why is that I knew it would produce several conditions and leave their effects, but I did not want any effects from past experience left on me and my sweet sixteen.

A girl will economize, go hungry and endure other hardships before she will put on another girl's dress to wear. I will put

6

gunny sacks on before I will wear another man's clothes, and this is only a step from having another fellow's girl or another girl's fellow.

Having such a case the present possessor would have to clean up the past performer's effects. Now you see, to clean up the other person's leavings, it is humiliating, so it would be a cheap and undesirable affair. I want one hundred per cent good or none. That is why I was so successful in resisting the natural urge for love making.

Now about sweet and how sweet, a girl can be one hundred per cent sweet to one only and no more. To illustrate, suppose we are two men and a girl together somewhere and some one else would ask if she is sweet and we both would say she is sweet. But let her act very friendly with the other fellow and then if someone should ask if she is sweet, I would say that she is not. Now you see her friendly action with the other fellow produced a change in me and it would

7

produce a change in any other normal man.

We always strive for perfection. We are only one-half of a perfect whole, man is the bigger and stronger half and the woman is the other. To be perfect there must be two, but where there is two there is no room for more, so the third party is left to go somewhere else with sour feelings.

A normal male is always ready to strive for perfection, the female is not. It is not only with human beings, it is the same with every living thing. If you watch a flock of chickens, where there is a rooster, and if you add another rooster, you will see them fight to death. One will have to go or be killed and this is the same thing with the other living things.

Lower forms of life are guided by instinct alone so the present only comes into consideration. As soon as the other male is chased away, the female is as good as she ever was, but with us it is different. We are guided more by reason and thought

than by instinct and so the present, past and future come into consideration. Now, if it is not good today, it was not good yesterday and it won't be good tomorrow. That is why an experienced girl cannot be one hundred per cent sweet.

According to my observation the girls are wrong in looking for their permanent partners. They are too quick. By being too quick, they only get those fellows with quick emotions. All quick emotions are irresponsible and short lasting.

There are two kinds of love—sensual and sentimental. Sensual love has the present and little future only. The sentimental love has the present, past and future, so it is more desirable. It will be slower but will last longer. Now, girls, when any fellow jumps quickly at you, you had better keep away from him. He is acting wholly selfish. He has no consideration that the action would do any good to you. You are the weakest side, so you

should have the better deal and if you don't get the better deal, there is a little brute in him and it may come very hard to train it out of him. The fellow who makes an advance toward you, and if he won't state what the eventual purpose will be, he is not a gentleman. All men should know that the girl's primary purpose is to find a permanent partner while they are young. Those fellows who fail to see this are not desirable to have around.

Girls below sixteen should not be allowed to associate with the boys, they are practicing in love making, such a thing should be discouraged. Love making should be reserved for their permanent partners. With every love making affair, their hearts get bruised and by the time they grow up, their hearts are so badly bruised that they are no more good.

Boys and girls start out as friends and finish as disappointed lovers, now let me tell you. Male and female are never friends, a

friend will not want anything from a friend, but a boy or a girl, one or the other, sooner or later, will ask for a little kiss, so they are not friends, they are lovers.

Let's see what happens when they are selecting their partners while they are young. They select their partners on account of good looks. The liking for the good looks remains but the good looks change and they change so much in ten years that you would not recognize them if you had not seen them now and then—and the boy gets the best deal almost every time. By that time they are grown up. The girls will be faded so much that the fellow would not want her any more so then, any girl who associates with a fellow only five years older is headed for a bad disappointment. This all could be avoided with the right kind of an education.

Now, a few words about education. You know we receive an education in the schools from books. All those books that

people became educated from twenty-five years ago, are wrong now, and those that are good now, will be wrong again twenty-five years from now. So if they are wrong then, they are also wrong now, and the one who is educated from the wrong books is not educated, he is misled. All books that are written are wrong, the one who is not educated cannot write a book and the one who is educated, is really not educated but he is misled and the one who is misled cannot write a book which is correct.

The misleading began when our far distant ancestors began to teach their descendants. You know they knew nothing but they passed their knowledge of nothing to the coming generations and it went so innocently that nobody noticed it. That is why we are not educated.

Now I will tell you what education is according to my reasoning. An educated person is one whose senses are refined. We are born as brutes, we remain and die as the

same if we do not become polished.   But all senses do not take polish.   Some   are   too coarse to take it. The main base of education is one's "self-respect".   Any one lacking self-respect cannot be educated.   The main bases of self-respect is the willingness to learn, to do only the things that are good and right, to believe only in the things that can be proved, to possess appreciation and self control.

Now, if you lack willingness to learn, you will remain as a brute and if you do things that are not good and right, you will be a low person, and if you believe in things that cannot be proved, any feeble minded person can lead you, and if you lack appreciation, it takes away the incentive for good doing and if you lack self control you will never know the limit.

So all those lacking these characteristics in their makeup are not educated.

Domestic

The foundation of our physical and mental behaviour is laid while we are in infancy, so the responsibility of our shortcomings rest upon our mothers and fathers, but mainly upon our mothers.

Today, I myself would be better than I am if my mother and father had known how to raise me and the same is true for almost everybody else.

At the first contraction in any part of your body, you will never notice any mark, but keep it up and some day you will see a crease, and it will be permanent. We all want to look and act the best that we know how, but we cannot learn from ourselves so we have to learn from others.

In my thirty years of studying conditions and their effects I have come to the conclusion that I can tell pointers to the people that would be a good help to them. That is why I wrote this little book.

To accomplish g o o d results, the mothers will have to keep good watch on their darlings until they acquire the natural ambition to shine, and the girls should be more carefully watched than the boys, because the girl's looks are her best asset and should be cultivated.

Don't raise the girls too big by over feeding them and too curved by neglect.

People who want to shine will always have to restrain themselves, because if they don't, their actions won't be graceful. Even when one's looks are good, if he abandons restraint, the performance won't be good. It is more likely that the person himself won't notice but others will.

The first thing I notice about other people is, if there is something wrong and if it could be improved and the same must be true about other people noticing my defects and neglects. To correct those neglects, somebody will have to point them out, but to do it directly will not do, because

15

they would think you are mean. That is why I want to point out the defects and neglects in this book.

The most striking neglect that comes to my attention is when one is smiling. A smile is always pleasing if it is regulated but without restraint, it is not. When smiling, the teeth only should be shown. As soon as you show the gums, it spoils the good effect. When showing the gums you are doing triple harm. First, the gums never look good; second, you are making too big creases in the side of your mouth and third, your lips come too wide apart. Especially should a girl be careful not to show too abnormally big mouth. Girls should do nothing that would impair their best looks. I have seen moving picture stars, public singers and others with their mouths open so wide that you would think the person lacks refinement, but if they knew how bad it looks they would train it out. No doubt they have practiced before

a looking glass, but a looking glass does not show such an enormous opening, because while they are looking at the looking glass they are under restraint and so they really don't know how it looks while they are not watching themselves. In a looking glass you will never notice all your neglects and defects. They have to be pointed out by somebody else.

It is painful to hear other people pointing out our neglects and defects so do not entrust your friends to do it. Your friends may not always be your friends. The best way is to leave that to your own family. Your mother and father will do nothing to embarrass you. Your mother will do it better and it should be started while one is still a baby.

The first thing the mothers should do is to watch the baby's mouth so it is not hanging open. The mouth, by hanging open, stretches the upper lip and when kept open while growing, then when fully grown,

the lips will not fit together any more.

Mothers should keep close watch on their children's behaviour. As soon as they notice some action and contraction that is not graceful, they should correct it immediately, because their actions leave their effects. To small children, it doesn't matter how ugly they look, but when they are grown up, the good looks will be the best thing, and one with a disfigured face cannot be satisfied with oneself. The foundation for one's best looks will have to be laid while one is small.

A graceful smile is pleasing but if it is not perfect, its pleasing effect is marred. To obtain better results, don't make the smile too big by opening the mouth too wide, drawing the lips over the gums, or drawing one side of the lip more than the other, or drawing both lips to one side and have them twisted.

Children should not be encouraged to smile too much, smiling in due time will

produce creases in the sides of their mouths. It would be better to save the smiles till they are grown up. Children while they are growing should be watched, closely. They are stretching their mouths with their fingers and are jamming too big objects in their mouths and making too ugly faces. All those actions should be forbidden for their future's sake.

Eyes should be trained to look in the middle between both lids, never through the forehead. If this is done, it will produce creases in the forehead. When the lids of one eye are more narrow than the other it should be trained out and equalized.

In case one leg is shorter and one shoulder lower, they can be disguised so that other people would not notice it. In walking the toes should be carried a little out, by carrying the toes out one can walk better. Shorter steps would make the walking more graceful and those who stoop over, higher

heels would help to keep the body more erect.

Girls should take smaller steps than boys. By taking smaller steps the body would not jump as much up and down or swing from side to side.

Mothers should study the other people's children and then pick out the best model from which to train their own child.

Everybody should be trained not to go out anywhere before somebody else has examined them to see if everything is all right. It would save many people from unexpected embarrassment.

Political

Before I say anything about the government, let's establish a base for reasoning. All our ideas should produce good and lasting results and then anything that is good now would have been good in the past and it will be good in the future and it will be good under any circumstances, so any idea that does not cover all this broad base is no good.

To be right, one's thought will have to be based on natural facts, for really, Mother Nature only can tell what is right and what is wrong and the way that things should be.

My definition of right is that right is anything in nature that exists without artificial modification and all the others are wrong.

Now suppose you would say it is wrong. In that case, I would say you are wrong yourself because you came into this world through natural circumstances that you had nothing to do with and so as long as such

a thing exists as yourself, I am right and you are wrong. Only those are right whose thoughts are based on natural facts and inclinations.

It is natural tendency for all living things to take it easy. You watch any living thing you want to, and you will see that as soon as they fill up, they will lie down and take it easy.

The physical comfort, the ease, that is the only thing in this world that satisfies. It cannot be overdone and it is the real base of all our actions. We all cannot take things easy because there is too much competition from other people only those who possess good management will succeed by exploiting domestic animals, machinery, other people and natural resources.

Everything will have to be produced that is consumed and to those who have to produce the things themselves, they are consuming the easy days are not coming to them.

It has been told to you that the government is for the purpose of protecting "life" and property, but it really is to protect "property" and life. Nobody wants your life but everybody wants your property.

In International dealings, when an army conquers the land, they don't want the people, they want the physical property and so do the thieves and the bandits. They want your money and property and if you will submit peacefully, they won't harm you.

Now you see, nobody wants you, they want your property so really the property is the one that needs the protection and not you. You are the protector yourself.

Government to be lasting will have to be just. This means it will have to protect all the property alike and all the property will have to pay equal taxes, which means big property, big taxes, and small property, small taxes. Government cannot exist without taxes so only those who pay taxes

should vote and vote according to the taxes they pay.

It is not sound to allow the weaklings to vote. Any one who is too weak to make his own living is not strong enough to vote, because their weak influence weakens the state and a degenerated state cannot exist very long, but every state should be sound and lasting.

By voting, the voters dictate the state's destiny for times to come and then to allow such a weak influence to guide the state, it is not wise and so you see one should vote according to how he is carrying the State's burden.

Another unwise thing about equal voting is that it gives the loafers and weaklings the power to take the property away from producers and stronger people, and another unjust thing about equal voting is that it gives the loafers and weaklings the power to demand an easy life from the producers and leaders.

Self respecting producers will not stand such an injustice for long. It is not the producers' fault when one is too weak to make his own living. The producer's life is just as sweet as the weaklings and loafers life is to them. All people are independent so you see everybody will have to take care of themselves and if they cannot, they should perish and the sooner they perish the better it will be.

To be lasting, the government should be built in the same way as the Supreme power of the land "the army." Governments have been rising and falling but the army always remains. You know there is no equality in army and so there can be no equality in the state if you are not equal producer you cannot be an equal consumer.

Fifty per cent of the people don't want to lead, they want to follow. They want somebody else to furnish the money for their living expenses and as long as such a condition exists, they are not equal with their

leaders. That is the reason why everybody should be put in the right place according to their physical and mental ability.

There is only one way to share the National income. It is by sharing the production and if you are not producing equally you cannot share equally. Nobody is producing anything for others. They are producing only for themselves.

People are individuals. For instance, if you want an excitement you will have to test the thrill yourself, or if you have a pain you will have to bear it yourself, or if you want to eat you will have to eat for yourself. Nobody can eat for you and so it is that if you want the things to eat you will have to produce them yourself and if you are too weak, too lazy, lack machinery and good management to produce them, you should perish and that is all there is to it.

STATISTICS ON THE CORAL CASTLE OF FLORIDA

The Coral Castle of Florida is the finest example of massive stone construction in the United States. A study of the immense sizes and weights of coral rock excavated, moved and used in its construction, in comparison with those used in many of the famous works around the world, establishes it as an authentic wonder of the world. The fact that it was accomplished entirely by one man makes it all the more remarkable.

Altogether there are approximately 1,000 tons of coral rock used in the construction of the walls and tower alone, a stupendous achievement for one man, unequalled in all history. In addition over 100 tons of coral rock were used in the carvings of the artistic objects throughout the entire park.

SOUTH AND WEST WALLS

There are 65 sections of stone weighing a total of approximately 420 tons, or an average of $6\frac{1}{2}$ tons each in these walls.

EAST WALL

There are a total of approximately 240 tons of coral rock in this wall which contains the following:

Crescent of the East . . .	23	Tons
Planet Mars	18	Tons
Planet Saturn	18	Tons
Obelisk	28½	Tons

NORTH WALL

Contains 149½ tons of stone. This wall contains the heaviest stone in the park, weight 29 tons; also the Polaris telescope, 25 feet tall and weighing 28 tons.

THE TOWER

The Tower contains 243 tons of coral rock made up of huge blocks of stone weighing up to 9 tons each.

The average weight of the stones used in the construction of the Coral Castle of Florida is greater than those used in the building of the Great Pyramid of Giza, while several of the stones used are taller than those found in the Stonehenge in England, and in weight exceed greatly the stones used in many of the other famous stone works throughout the world.